I0471170

Public Fire Education Planning for Rural Communities:
A Five-Step Process

Sharon Gamache
John R. Hall, Jr.
Marty Ahrens
Geri Penney
Ed Kirtley

October 2007

Provided for by Cooperative Agreement EME-2004-CA-01867

Table of Contents

Introduction 1

Step 1: Conduct a Community Analysis 6

Step 2: Develop Partnerships in the Community 16

Step 3: Create a Strategy to Solve the Problem 23

Step 4: Implement the Strategy in the Community 32

Step 5: Evaluate the Results 38

Summary 43

Introduction

Congratulations! You are about to take the first step in making your community fire safe. But you don't have to go it alone—you also are partnering with numerous Federal and nonprofit agencies and organizations committed to helping you in this project.

This planning guide will introduce you to a simple, easy-to-use planning process to develop a fire safety program for your community. Since the planning process is based on the experiences of other rural communities, we know all the steps work. Examples throughout the guide will help you apply each step to your community. These examples, along with the other information, make using the steps straightforward. All of the information in this planning guide is based on the U.S. Fire Administration's (USFA's) booklet *Public Fire Education Planning: A Five-Step Process.*

You can use this planning process in any rural community in the United States, regardless of the community's location, size, population demographics, or fire problem. The process leads you to solutions based specifically on your community's resources, problems, and values. It is **not** a large-city process that requires large numbers of firefighters and thousands of dollars to be effective.

Even though the process is designed to work for any rural community, we have made some assumptions about your situation:

- First, only a few people in your department may have experience developing a fire safety education program for the community. You may have personnel who have done presentations during Fire Prevention Week or who help with fire station tours, but they may not have designed a comprehensive public education program.

- Second, you don't have many available fire prevention resources, including people's time. You may have a hard time even getting enough qualified people on the department to fight fires. Also, you may not have many department funds available for a fire prevention program.

- Third, you may only have a small database of fires in your local community. Your department probably does not have many structure fires each year. Because of the low number of fires, there may not be any single fire cause that you can single out with confidence as a top priority.

Some elements are common in most successful rural fire safety education programs. Knowing these common elements is important as you work through the planning process.

- **Individuals and the department are committed to the fire safety education program.** Your interest in developing a fire safety education program for the community is an example of this commitment. You will most likely be the "sparkplug" for the program in your department, which will take some dedication on your part and will require a high level of commitment—from beginning to end. In addition, the leaders of the department—including the fire chief—must understand that the program will require the commitment of the department. This commitment may be in firefighter time, in some funding, or in helping you get other members of the community involved.

- **The fire safety education program is based on the local fire problem and the people who are most likely to be victims of fire.** Few, if any, successful fire safety education programs are put together without any planning. Action without planning is likely to lead to more expense and less safety improvement. Rural communities and small fire departments with successful programs have used a planning process to identify the most critical problem and develop and implement a solution.

- **Partnerships in the community are established so that the community is involved in the planning process and the solutions to the problem.** Because your departmental resources are limited, you'll need to work with other agencies and organizations in the community. This partnership approach gives your program more resources and greatly increases your chances of success. Also important, the teamwork approach builds the partners' vested interest in the program's success.

- **The fire safety education program is evaluated in some manner to determine if it is successful.** The evaluation of the program does not necessarily mean a complicated and costly statistical process that takes forever to complete. You can do a great deal of valuable evaluation in simple and straightforward ways. Regardless of the type of evaluation used, something is done to identify what worked, what didn't work, and what to do next time to improve the results of the program. The evaluation also may provide you with the ammunition you need for additional funding and resources in the future.

Again, congratulations on your decision to improve your community's fire safety. While you have considerable work ahead of you, you can be successful. Section 2 provides an overview of why you should use this planning process to design your fire safety education program. Good luck!

Why Use a Planning Process?

Firefighters are trained to take action—it is a part of the fire service culture. In emergencies, firefighters respond quickly to solve the problem, and it works well. When implementing a fire safety education program, the temptation to "just get something done" is hard to resist. Yes, scheduling a few presentations at a local elementary school during Fire Prevention Week, passing out brochures at a county fair or community event, and giving stickers and plastic helmets to children during a parade are all easy. But do these activities really improve the fire safety of your community? Do they convince citizens to take action, such as installing a smoke alarm? Are you reaching the highest risk groups? Unfortunately, the answer to these questions is sometimes "No!"

The business-as-usual approach will not work. Seldom does it really reduce the actual fire risk in the community. Regardless of the size of your community or fire department, however, you can have success using a simple, proven five-step planning process.

Step 1: Conduct a Community Analysis

The community analysis will identify the fire problems in your community by using information from your fire department, other agencies in the community, and a few State and national resources to understand your community's highest priority fire problems.

Step 2: Develop Partnerships in the Community

A community partner is a person, group, or organization in the community who will join you to reduce the fire problem. The most effective fire safety education efforts are those that involve the community in the planning and implementation of the program and place a strong emphasis on maximizing the available resources in the community.

Step 3: Create a Strategy to Solve the Problem

The strategy is the solution or solutions to the problem and constitutes the beginning of the detailed work necessary to development the fire safety education program.

Step 4: Implement the Strategy in the Community

Implementing the strategy involves putting the plan into action in the community. The implementation step must be well coordinated.

Step 5: Evaluate the Results

The primary goal of the evaluation process is to demonstrate that the fire safety efforts are doing what you intended. In other words, the evaluation step is about determining if the program achieved its goals. You will use the information you gather in this step to help you develop the next fire safety education program.

Consider what is accomplished at the end of each step.

Step 1: A clear understanding of the community's fire problem and the people affected by the problem.

Step 2: The people and organizations in the community who are concerned about fire safety come together to develop a solution to the problem.

Step 3: A plan is developed with solutions most likely to reduce the fire problem.

Step 4: The plan developed in Step 3 is implemented as designed.

Step 5: Information from the program is used to determine if the program accomplished its goals.

This process has several benefits. First, you will identify the fire problem clearly. While you probably have a general sense of your community's fire problems, the planning process helps you address the highest priorities. Also, you will have a better understanding of who is involved, specifically those citizens who are most affected when a fire occurs.

Another benefit is that you'll maximize community resources. Usually no single agency or department has enough resources to address the fire problem effectively. Success requires bringing the various community resources together to focus on the solutions.

The third benefit is the community involvement. Your fire problem is actually the community's fire problem. When a fire occurs, the effects are felt throughout the community: individual families, insurance companies, schools, local government, local businesses, churches, and so on. Because the problem belongs to the community, the community should have a stake in the solutions.

The next benefit actually involves future fire safety initiatives. By using the five-step process everyone involved will develop skills and experiences that can be used in future community projects, including those not related to fire safety.

The final, and most important benefit, is that when you and the community use the five-step process the likelihood of success is greatly increased. You will achieve the program goals and, in the end, the citizens of your community will be safer from fire.

While you will see many benefits from the five-step process, be aware of a few challenges, which do not take away from the importance of the process, but rather are issues for you to consider as you and your community team work through the five steps.

Successful use of the five-step process requires a commitment by those involved. If your goal is simply to have the fire truck in the parade once a year, you don't need the five-step process. If, however, you want to reduce the fire problem in the community, you will need to invest some time and effort. The worst thing that can happen is to get part way through the process and find that those involved have given up or lost interest. Anyone that agrees to be involved, regardless of the level of involvement, must make a commitment to carry through.

You need strong support from fire department leadership. Strong support from the department's leadership will ensure that department members are involved and supportive. In the largest and smallest fire departments alike, some fire service members may resist change, especially if they perceive that the fire safety education program may divert limited resources from suppression activities.

You may need assistance from outside resources during the planning process. In some cases, often during Step 1, the planning team may have to seek help from outside agencies. You have nothing to lose by asking for help from individuals and organizations experienced with a five-step planning process.

The community may not have a single, significant fire problem. Sometimes, getting citizens and community leaders excited about a problem or risk they perceive as a "non-issue" is difficult. Many rural communities experience only a few fires each year. This doesn't mean, however, that the fire risk does not exist, especially to the very young and the elderly. Also, people who live in rural areas may be at greater risk due to longer response times, limited water supplies, fewer governmental services, and so on. Fortunately, some fire risks are common to every community and always need to be addressed, such as homes that lack working smoke alarms.

Getting the community involved in your fire safety program has two tangible benefits to your fire department. Through the five-step process you will develop close working relationships with community leaders and citizens who have influence and access to many kinds of resources. These relationships are invaluable when applied to other fire department issues, such as fundraising, recruiting, and so on.

Also, through the five-step process you will create or earn credibility with the citizens in the community. Although most fire departments already enjoy community respect, the process enhances that respect and credibility. As with relationships, this credibility is invaluable when seeking community support for other activities and initiatives.

The five-step planning process is a proven method for addressing a community's fire problem that can be used in any community. Both the community as a whole and the fire department will enjoy numerous benefits from using the five-step process. Success does, however, require commitment by the department and others involved.

Step 1: *Conduct a Community Analysis*

A community analysis identifies and prioritizes the fire problems in your community and provides information on those citizens affected by the fire problem. The analysis sets the foundation for developing your community fire safety education program. Most likely, you do not have the resources to fight every fire problem, so focusing your efforts will be important. Tackle the worst problem first, but keep the analysis process ongoing since the fire situation may change over time.

Take the time to conduct a community analysis for several reasons. First, you need facts to make good decisions. You may have some ideas about the problem, but a community analysis will help you be certain. The more you understand the community's fire problem, the easier developing and implementing the correct solutions will be.

The term "fire problem" encompasses several elements, including actual fires and their known causes. Another part of the fire problem includes common hazards, even if no fires result from these hazards. "Hazard" means a condition or situation with the potential for unwanted consequences. "Hazard" need not imply the object in question is inherently unsafe. All types of heating equipment inside the home can be fire hazards, for example, if their condition, location, or manner of operation do not comply with rules of safe operation, but that does not mean that the heating equipment is generally unsafe.

Know where your community stands before beginning your fire safety education program. Too often, fire departments overlook this step and buy a program that looks good to address a fire problem that may not even exist. Other times, a fire safety educator will pick a topic to teach simply because he or she is interested in the topic or teaching the topic is easy.

Don't become distracted from local issues by media coverage of national events. Stay focused on your community's needs. Invest the time to conduct an objective community analysis. Having the facts about your community's fire problems is the first step to a successful fire safety education program.

A community analysis includes five important activities:

1. Identify available information about your local fire problem.

2. Develop a community fire profile.

3. Write a problem statement.

4. Prioritize the fire problems.

5. Identify the people affected by the fire problems.

The analysis provides a factual overview of the fire problem, using available information that you can use to develop a picture of what is wrong and who is at risk.

1. Identify Available Information about Your Local Fire Problem

Information about the fire problem identifies the leading causes of fire, where the fire problem is occurring, and who the fire problem is affecting.

Obtain problem-related information by conducting research, asking questions, and reviewing information. Start with local information, then review State information, and finally review available national information. In other words, begin your work with information available in your community, because that local information most likely will paint an accurate picture of your specific fire problems.

There are numerous sources for each level of information, starting with local information. Remember that the State and national information generally is developed from local information from each community in the State and region.

Your department's incident reports are the best source of local information on your community's fire problem. The incident reports are a lasting record of the fires to which you have responded, the people affected, the cause, and the resulting damage. Some rural fire departments computerize their reports, while others use paper reports and a filing system. Most U.S. fire departments use a standardized form from the USFA's National Fire Incident Reporting System (NFIRS). Most States also use this form, or a variation of it, to record fire incident information. You can get more information on NFIRS at *www. NFIRS.FEMA.gov*

Review the NFIRS form or your department's incident reporting form. Several pieces of information on the form are important to your analysis in terms of seeing your local fire problem clearly. To identify patterns, choose an extended period—at least 4 or 5 years—to use as a baseline, that is, as a representative description of your fire problem just before the introduction of your new safety program(s). If your department participates in NFIRS, you should be able to generate standardized output reports on any fire or victim characteristic. If not, develop a tally sheet for each of the characteristics of interest, which may include

- location of the fire (block or neighborhood, property use, area of origin);

- cause of ignition (heat source, equipment involved in ignition, item first ignited, factor contributing to ignition);

- human factors contributing to the fire or to a victim's injury or death; and

- damage from the fire.

Once you have reviewed the information in the incident reports you should begin to understand

* the area of the community or county where fires occur most frequently;

* the cause of the fires;

* when the fires occur—time, day, month;

* any contributing human behavior;

* the presence or absence of working smoke alarms;

* the amount of damage caused by the fires; and

* people affected by the fire.

If you are using tally sheets and, even with several years of data, have very few incidents to work with, treat each fire like a case study without leaning too much on statistical methods. Take notes from each section you believe is important. One method is to list the cause, source of ignition, and so on. If you come across another incident with an already noted cause, for example, put a check by it. This will show you which causes, human factors, addresses, and so on are occurring more frequently. In other words, you will begin to identify patterns. Also pay attention to information in the report's narrative, called "anecdotal" information. This information, while not necessarily statistically representative, can provide the reporting officer's perspective on what was most important in the incident and, in particular, will help provide details that do not fit well with the choices in the coded data fields.

You also should review other local sources of information, some of which are listed below. Remember, though, that specific privacy laws affect the release of information about fires, burn injuries, and legal investigations. You will have to work with local agencies and officials to get the needed information without violating privacy laws.

Insurance agents. Insurance companies keep reports of every fire for which a claim is flied. Local insurance agents generally have access to this information. Also, insurance agents may be able to provide information on fires in nearby communities. Finally, insurance agents may provide anecdotal information on the affected families.

Hospital reports. Hospital and emergency medical services (EMS) reports can provide information on any burn victims, including the cause of the burn.

American Red Cross reports. In most communities, the American Red Cross (ARC) helps fire victims. Visit with the local ARC coordinator for information about families affected by fire.

Experiences of the firefighters. This source of information is important, especially with regard to the cause of fires.

Fire investigation reports. These reports, similar to fire incident reports, give more detailed information about the specific cause of the fire. In some cases, the local law enforcement agency or State Fire Marshal investigates fires.

News stories. Usually, the local newspaper covers residential fires that result in loss. Copies of the news stories should be available at the newspaper or the local library.

Health Department. In some communities, the health department is actively involved in fire and burn prevention activities. In these communities, the health department officials may already have gathered information about the local fire problem.

The next level of information comes from sources at the State level, many of which are similar to local sources. In general, information from these sources is less specific than local sources. The information, however, is still valuable and adds to the overall picture.

State Fire Marshal's Office. Each State has an office or agency that acts as a State Fire Marshal. One of the roles of the State Fire Marshal is to gather information on the State's fire problem. In some cases, the office puts the information into detailed reports sorted by county. In other cases, you must compile the information. Keep in mind that the fire marshal's information comes from reports submitted by each community in the State—a good reason for your department to complete and forward its own incident reports.

Insurance Commission. The State Insurance Commission may track types of claims, which may include causes of fires, amount of claims paid, and so on.

State Health Department. The health department will have information on burns, as well as on the cause of residential fires.

The final level of information comes from national sources, though it tends to be more State and regional in nature, as well as more general. Just as with State information, national information will add to the overall fire picture in your community. Most importantly, national information can provide insight into the fire problem in similar communities in your State and region.

National Fire Data Center (NFDC). The NFDC is part of the USFA. The NFDC collects, analyzes, publishes, disseminates, and markets information related to the Nation's fire problem and USFA programs. You can access all of the NFDC information on the Internet at *www.usfa.dhs.gov/statistics* One of the most helpful reports is a series called Fire in the United States, which provides information on the causes of fires, the people affected by fire, and the fire problems in the various regions of the country. Figure 1 shows a chart taken from the section on leading causes of fire.

National Fire Protection Association (NFPA). Every year, the NFPA publishes dozens of reports on the fire problem in the United States. These reports, which are available at *www.nfpa.org* are similar in topics, methods, and data sources to those of the NFDC and focus on detailed characteristics from a national perspective.

Centers for Disease Control and Prevention (CDC). The CDC is involved actively in reducing the U.S. fire problem. In fact, CDC has created several new fire safety programs. Figure 2 shows a sample of the information available at *www.cdc.gov/health/fire.htm*

Figure 1. Trends in Causes of Residential Fires and Fire Losses

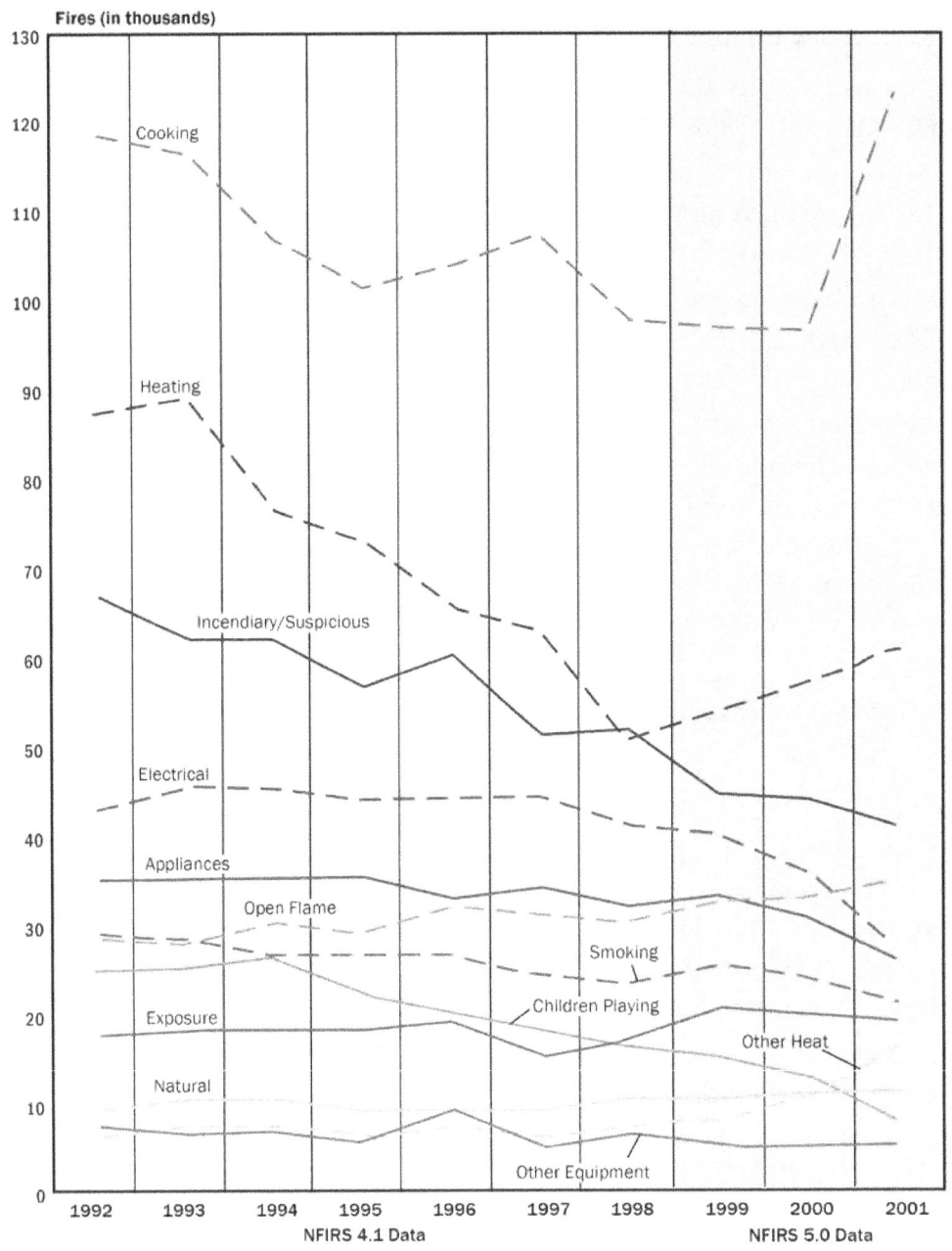

Note: Data provided in Appendix B, Table B-3.

U.S. Census. The U.S. Census Bureau offers significant demographic information on each community in the United States. While you may have a sense of your community's demographics, the census information provides details about who lives there, income levels, housing, and so on. Figure 3 shows census information for Boise City, Oklahoma. You can find information on each community in the American Fact Finder section of the census Web site at *www.factfinder.census.gov*

Figure 2. CDC Fact Sheet on Fire

SAFER · HEALTHIER · PEOPLE

Fire Deaths and Injuries: Fact Sheet

Overview

Deaths from fires and burns are the fifth most common cause of unintentional injury deaths in the United States (CDC 2005) and the third leading cause of fatal home injury (Runyan 2004). The United States' mortality rate from fires ranks sixth among the 25 developed countries for which statistics are available (International Association for the Study of Insurance Economics 2003).

Although the number of fatalities and injuries caused by residential fires has declined gradually over the past several decades, many residential fire-related deaths remain preventable and continue to pose a significant public health problem.

Occurrence and Consequences

- On average in the United States in 2005, someone died in a fire about every 2 hours (143 minutes), and someone was injured every 29 minutes (Karter 2006).
- Four out of five U.S. fire deaths in 2005 occurred in homes (Karter 2006).
- In 2005, fire departments responded to 396,000 home fires in the United States, which claimed the lives of 3,030 people (not including firefighters) and injured another 13,825, not including firefighters (Karter 2006).
- Most victims of fires die from smoke or toxic gases and not from burns (Hall 2001).
- Smoking is the leading cause of fire-related deaths (Ahrens 2003).
- Cooking is the primary cause of residential fires (Ahrens 2003).

Figure 3. 2000 Census Information for Boise City, Oklahoma

General Characteristics	Number	Percent	U.S.
Total population	1,483		
Male	704	47.5	49.1%
Female	779	52.5	50.9%
Median age (years)	41.1	(X)	35.3
Under 5 years	105	7.1	6.8%
18 years and over	1,095	73.8	74.3%
65 years and over	314	21.2	12.4%
One race	1,441	97.2	97.6%
White	1,211	81.7	75.1%
Black or African-American	3	0.2	12.3%
American Indian and Alaska Native	25	1.7	0.9%
Asian	4	0.3	3.6%
Native Hawaiian and Other Pacific Islander	0	0.0	0.1%
Some other race	198	13.4	5.5%
Two or more races	42	2.8	2.4%
Hispanic or Latino (of any race)	312	21.0	12.5%
Household population	1,442	97.2	97.2%
Group quarters population	41	2.8	2.8%
Average household size	2.36	(X)	2.59
Average family size	2.99	(X)	3.14
Total housing units	752		
Occupied housing units	610	81.1	91.0%
Owner-occupied housing units	449	73.6	66.2%
Renter-occupied housing units	161	26.4	33.8%
Vacant housing units	142	18.9	9.0%

2. Develop a Community Fire Profile

A community fire profile is an overview of the information previously gathered about the community and can serve as an introduction to a problem statement. A community fire profile should include the following information:

- a description of the demographics of the community;

- a brief description of the fire problems;

- the public perception of the fire problem, which may be based on the experiences of firefighters;

- the level of community support for fire safety education; and

- community resources available to address the fire problem.

A community fire profile doesn't have to be long—a paragraph or two for each section is enough. The community fire profile leads a reader to the problem statement. Include one profile description for each individual fire problem and provide adequate information to give readers a basic understanding of the community's fire problem.

Community Fire Profile for Small Town

Small Town, USA, has a population of 1,600 and is in the Midwest. The town grew from a small village to its present size in the early 1900s. Most homes are more than 40 years old and constructed of wood. Although farming once thrived in Small Town, most residents now commute to a large city 50 miles away.

Small Town is an aging community, in terms of people and property. A significant 40 percent of Small Town's citizens are retired and reside in older homes. A large portion of the retired residents have difficulty with mobility and live on a fixed income.

The local fire chief's analysis of the fire problem identified that gas wall heaters were involved in the majority of fires in the community. In fact, in the past 3 years, malfunctioning wall heaters or combustibles too close to the heaters caused 11 of 15 structure fires. Of those 11, 8 were in the homes of retired residents.

A recent survey by the county health department showed that only 25 percent of the homes had working smoke alarms. Also, the survey determined that only 10 percent of the citizens knew that unattended cooking was a leading cause of fire.

3. Write a Problem Statement

A problem statement provides a fact-based overview of the problem and who it affects. It also provides the fire department's vision in terms of a solution. The problem statement does not have to be long—it just needs to give an objective and accurate description of the fire problem.

Share your problem statement with the community and convince others to join the team! A poorly written problem statement that lacks facts may damage your fire department's credibility and result in less support for your proposed fire safety education program.

Remember that you'll use the problem statement as a marketing tool when you ask others to become partners in your program and when seeking resources. Because of its importance to the development of the program, taking the time to develop a well-written problem statement is important.

Small Town Problem Statement

Wall heaters cause 75 percent of the fires in our community. More than half of the residential fires involved retired citizens who live in older homes. Most of the unattended fires are due to a heater malfunction, or combustibles too close to the heater. Also, only one in four homes in Small Town have working smoke alarms.

The Small Town Volunteer Fire Department proposes developing a fire safety education program to address the wall heater fire problem among retired citizens. The reasons for this fire safety education program include

- *A majority of the residential fires that occur in Small Town involve gas wall heaters.*
- *These fires affect our retired senior citizens, a group at highest risk from fire.*
- *Seventy-five percent of the homes in our community do not have working smoke alarms.*
- *A fire safety education program can reduce the fire problem involving senior citizens in Small Town.*

4. Prioritize the Fire Problem

Your community may have several fire problems. Select one problem as the priority and work to reduce it before moving on to the next. Making this choice, which you should base on local needs, takes time, effort, and patience. When deciding the highest priority fire problem, consider the number of citizens killed or injured by fire; the amount of dollar loss caused; and any rapid increase in the number of fires.

The most frequent fire cause may not always be the highest priority. In your community, for example, the most frequent cause of fire may be careless burning of brush and trash in the rural areas. In the course of 3 years, perhaps your department responded to 30 such fires. The structural damage was minimal, however, with no injuries or deaths. During that same period, assume you had two structure fires caused by young children playing with matches. One of the fires left two children hospitalized with burns and one of the houses was a total loss.

Which of these problems is the highest priority for your community? If you made the decision based only on the most frequent type of fire—brush fires—you would miss the type of fire with the severest impact on your community's citizens. In short, consider each fire problem based on its overall impact to the community and its citizens. The best choice—the highest priority fire problem—is generally the problem with the highest human impact.

5. Identify the People Affected by the Fire Problems

Certain citizen groups suffer a much higher rate of fire risk. Factors that can increase this risk from fire in different groups include

- age, particularly children under 5 and adults over 65;

- developmental, physical, visual, and hearing disabilities;

- sociocultural and economic status;

- gender, with males historically having higher rates of fire; and

- language and communication barriers.

A disability, which can affect any group in the community, can affect a person's ability to react and respond effectively to a fire. All communities are socially and culturally diverse. Changing family structures, peer influences, and language diversity are all examples of social and cultural issues you must consider.

Invest the time and effort to learn about your community. The success of your fire safety education program will depend on finding out who is most at risk from each of the high priority fire problems. Failing to learn about the "people aspect" of the fire problem can result in future programs not reaching the right groups.

Making assumptions about high-risk groups can be dangerous. Several national organizations, including NFPA and the USFA, have researched the most common high-risk groups. Review the research to learn more about what fire hazards pose the most risk to each group. This information will be critical as you develop your solutions.

Summary

A community fire analysis identifies the fire problems in your community, and helps you to understand the various groups who live in the community and are most affected by fire. The information developed from the analysis creates the foundation for developing your fire safety education program.

Step 2: *Develop Partnerships in the Community*

Community partnerships and networks are essential for successful mutual-aid relationships. The same is true when attacking your community fire problem. No matter your effort and commitment, making a positive change without help from others will be difficult. Having multiple partners or local networks makes sense in terms of increasing resources to tackle the problem, which includes identifying the fire problem, developing a plan, and implementing the solutions. The most successful fire safety education programs involve the community in the planning and solution processes.

Developing partnerships involves the following activities:

1. Identify possible community partner.
2. Form a community planning team.

1. Identify Possible Community Partners

Think teamwork! As with firefighting, community fire safety education is best done by a group. Working alone will consume a great deal of time and resources and seldom is effective in reducing the community's fire problem. The most effective and time-wise approach is to build a partnership around an existing network.

When people hear "resources," they immediately think of money. While financial funding certainly is important, money alone will not accomplish everything. Generally, the needed resources fall into four categories: wealth, work, wisdom, and influence.

Wealth. Wealth, of course, is about funding. Every fire safety education program requires funding. Money is needed for materials, events, and purchasing equipment, such as smoke alarms. At times, funding will come from the department's budget. Other times, you may have to go to the community for funding.

In-kind support is a type of wealth that involves resources in lieu of money and may include equipment, printed materials, supplies, and personal time. Donated professional services, such as consultations, program evaluation, and design of educational materials, are all examples of in-kind support.

Work. Work involves the people actually conducting the fire safety education program. If, for example, you plan to install smoke alarms in the homes of older adults, you will need people who can install the alarms. Often, members of the department will do the work. Other times, you may have to recruit help from the public, other community organizations, or service groups, or you may have to create an organization.

Wisdom. Wisdom involves the skills to plan, develop, implement, and evaluate the fire program. If you are lacking in this area, reach out to others. You may, for example, not have any experience evaluating a program. Perhaps your County Extension Agent is experienced in this area and can guide you through the planning and evaluation. This is a case of using someone else's experience and skills to support the program.

Influence. Influence involves persuading others to take action or get involved. In a rural community, a few people probably always seem to be in the center of what's going on—those you might call the "key influential people." Think about who these people are in your community and set your sights on signing them up for your program. Where they go, others will follow. One type of influence is political support. Support from elected officials for fire safety education is crucial. Discuss any plans or ideas with elected officials who have a stake in the program's success. This may be a city council, district fire board, fire commission, or county commissioner. These people may know how to obtain additional resources and often will have information important to the program.

Identifying a group of partners who have a stake in the success of the fire safety education program is important. Many people and groups can offer insight into the people who live and work in the community and the fire problems they experience. A community partner is a person, group, or organization willing to join you and address the fire problem. Often, the partners will depend on the specific fire problem. As a leader of the fire safety education program, you should identify who can bring work, wealth, wisdom, or influence to the program and get them involved.

Defining the concept of a community network is important, as well. A network, as used here, consists of a central organization with existing relationships for a specific purpose in the community. In some communities, the central organization may be the fire department. In other communities, however, the central organization may be the county health department, faith group, school, agency on aging, senior citizen center, Cooperative Extension, or other government agency. The identity of the central organization is not as important as whether it has an established network in the community that can be used for fire prevention initiatives.

At this point, you should have selected the highest priority fire problem. Think about which groups in the community the fire problem affects. Consider which organizations or agencies already provide services to these people, and who cares about the specific fire problem and the people it affects. These people may be the best to be involved in the fire safety education program. Another approach is to find out who has the resources needed to address the problem. You may determine, for example, that your community's highest priority fire problem involves older adults. If your community has an active senior citizens' group and several of the local churches have frequent events for elderly citizens, the churches and senior citizens' group should be your partners.

Build around existing networks. Start building a team of partners by contacting the people you know in the community who are part of well-established networks, who are well-known, well-established, and trusted in the community. Explain the program and what you hope to achieve. They may be able to identify potential partners. In addition, use the Internet and telephone directory. Find out which groups might be interested in a specific fire problem or that already address fire problems. Throughout this step, avoid duplicating the effort of other agencies and organizations.

Possible community partners:

- groups already interested in addressing the fire problem or similar safety problem;

- health departments, public health personnel, and individual private care providers;

- churches and other faith groups;

- schools;

- rural electrical cooperatives;

- members of the community the fire problem affects;

- cooperative extension;

- people or groups that feel the financial impact of the fire problem, including insurance companies, property owners, ARC, and so on;

- groups or organizations that provide services to the affected groups;

- community service and advocacy groups; and

- groups that can help deliver fire safety messages, including the media.

While working alone may seem like a good idea; failure to obtain the perspective and input of others in the community will limit the success of your program. Recruit a core group of people with an interest in solving the fire problem—called stakeholders. These stakeholders will share the responsibility for developing and implementing a quality fire safety education program.

Don't expect everyone in the community to support your fire safety education program instantly. Many may not even believe that a fire problem exists. Instead, expect them to be part of the solution by educating them about the problem. Your community profile and problem statement are powerful tools, justifying why others should partner with you. Share your vision for the program and possible solutions with community leaders.

The most effective fire safety education programs have community "buy in" by involving the community in the planning and solution processes. The community must understand that a problem exists and that it can be solved. Sharing your collected information in a professional manner will be important. Be sure to explain how you obtained the information and from where. Then work with others to continue building the program.

2. Form a Community Planning Team

Discuss your fire department's intention of forming a planning team to tackle the community's fire problem with the leaders and community organizations you identified as potential partners. If possible, try to meet individually with each group or person;

meeting individually is more personal, and allows more time to answer specific questions. Be prepared to explain what resource that person, group, or organization can provide. Also, be prepared to answer questions about what you expect from them. The more indepth you can be, the more likely you will gain their cooperation. When you meet with the person or group to ask if they will join the planning team, solicit suggestions about other potential partners and possibly changing your approach.

Once you have assembled a planning team, schedule a 1-hour team meeting with an agenda. Your partners will expect you to use their time efficiently. If needed, ask someone with experience working with groups to help you put together the agenda and to facilitate the meeting.

Often community planning teams lead to the development of formal coalitions. A coalition is a group of people with different interests who come together to work on a common problem, such as the local chapter of National Safe Kids Coalition, the local chapter of Mothers Against Drunk Drivers, and neighborhood associations. Coalitions usually include representatives from several different groups in the community and lend themselves well to the advancement of fire safety education programs.

A Community Success Story: Johnson County (AR) Fire District #1

The Johnson County, Arkansas, Rural Fire District (RFD) #1 is a good example of how one small department worked with community partners to expand its fire prevention education programs. In 2002, the 19 members of the RFD had no fire prevention education budget but managed to provide about 30 hours per year of programming. In 3 years, RFD expanded to 115 programs and activities reaching nearly 25,000 people.

How did they do it? The department's success resulted from its relationship with Fire Corps, a program that uses members of the public to supplement fire department non-operational activities. Students from the University of the Ozarks had been helping the department with its fire safety program since 2002. But when RFD received its third grant in 2005, the firefighters realized they needed the students even more.

The department joined the national Fire Corps organization and asked the university if it would take on the department's program as a community service project. The students brought new skills and energy to the program, including more structure and organization, and helped with strategic planning and marketing. Program leaders worked through the summer to prepare for student-helpers arriving for the 2005-2006 school year. The group continues to be an important part of the department's programming, helping to reduce the district's property loss and burn injuries. In fact, in the first year, residential property loss from fires decreased by 34 percent while the number of burn injuries remained steady. As fire safety programming continues to increase, the amount of property damage in the district continues to decrease.

continued on next page

By partnering with Fire Corps, the department increased its fire safety programming in the first year by 287 percent, presenting more than 46 fire safety programs to 2,570 children and 624 adults and being involved in 69 other fire safety related activities that reached 24,817 people. The department's Fire Corps partners helped with 143 hours of presentations and activities. RFD's success has attracted recognition at the local, State, and national levels. The department won the Carnahan Award, presented by the Arkansas Fire Prevention Commission at the Arkansas Firefighter Convention. The department could not have accomplished what it did without the community partnership.

Source: http://www.rfd1.com/firecorps.html

Many programs and partners have been successful in rural communities, some involving fire safety and others involving similar issues. The U.S. Department of Homeland Security (DHS) highlights the following examples in its report "Mitigation of the Rural Fire Problem—Strategies Based on Original Research and Adaptation of Existing Best Practices."

USDA Loans and Grants[1]

The U.S. Department of Agriculture (USDA) Home Repair Loan and Grant Program (Section 504) offers low-interest loans of up to $20,000 and, for people who are at least 62 and unable to repay a loan, grants of up to $7,500 for repairs or the removal of health or safety hazards. Loan recipients also may use the funds for improvements or modernization. The USDA, through its community facilities program, also coordinates the Rural Emergency Responders Initiative, which offers financial assistance for equipment, vehicles, and buildings for fire, police, heath care, and other activities. The USDA gives priority for grants to low-income communities and communities with populations of less than 5,000, as well as to health care, public safety, and community or public service projects. The USDA also offers loans for rural areas and small towns with populations of up to 20,000.

Alaska's Micro Rural Fire Department Program[2]

Safety professionals in Alaska developed Project Code Red, or the Micro Rural Fire Department, to address the fire problem and lack of firefighting equipment in the State's small rural areas. This project uses new and existing technologies and State-certified training to cope with the extreme winter temperatures, lack of hydrants, and, in many cases, lack of roads. The program equips five firefighters with fully supplied trailers that they can pull by all-terrain vehicles, by snow machines, by pick-up trucks, or by hand, even on boardwalks and trails. Program facilitators ship the trailer in a heated and insulated container that doubles as a firehouse and includes 600 gallons of environmentally safe firefighting foam. Firefighters can recharge the unit in less than 5 minutes for less than 50 dollars. Also included in the shipment is State-certified firefighter training, based on an adapted version of NFPA Firefighter I, which is geared to fire departments without protective gear that have a limited water supply and may have only portable extinguishers and pumps. The total cost is about 70 percent less than the cost of a new fire engine.

African-American Churches in Rural Communities[3]

Stephanie Boddie discusses the role of African-American churches in rural communities and particularly in the largely African-American town of Boley, Oklahoma. Historically, many congregations have provided some type of social services. Boddie references Chaves and Higgins, who found that black congregations were more likely than white congregations to be involved in civil rights and providing basic needs to the immediate community. In fact, they describe rural churches as similar to an extended family and important forces in leadership development and community organizing. The churches also interact strongly with the private sector. In both rural and urban African-American communities, the churches are among the chief sources of influence and support.

Church leaders often provide members, leadership, and resources to social programs independent of government policies and sponsor a variety of programs, such as literacy, drug and alcohol prevention, and youth outreach. Fourteen of Boley's congregations have formed a ministerial alliance, and two of the three remaining assist when needed. The alliance "has been most effective in maintaining and institutionalizing the programs initiated by the various participating churches, including a community choir, a senior center, funeral ushers, a crime watch group, financial assistance, historic preservation, hosting of celebrations and meetings, sponsorship of the Sunday School Institute, vocational training, a volunteer fire department, and a literacy program that culminates in a GED."

Community Program at a Rural Manufacturing Plant[4]

Fires, Ripley, Figueiredo, and Thompson describe organizing efforts in smoking cessation and dietary improvements at a rural manufacturing facility in Mecklenburg County, Virginia. The authors describe the population, of which 80 percent live in rural areas, as "underserved in terms of medical care, patient education, and cancer education."

The project began by recruiting a 12-member Health Advisory Board (HAB) from among employee volunteers. The HAB's goal was to "identify and develop activities and health promotion ideas to address cancer prevention and to change diet and smoking among their co-workers."

With support from the project, the HAB conducted one activity per month over 9 months. Before the program, 40 percent of the survey respondents were smokers. After the program, the percentage dropped to 35. A "stop smoking" contest attracted 13, or 18 percent, of the known workplace smokers. All 13 quit for 24 hours and 5 were still not smoking after 2 months. In addition, smokers at the plant tend to be more ready to quit smoking at the end of the program than they were at the outset.

Not counting the donations of pamphlets, posters, and prizes or the employees' time, program costs totaled about $1,925. The authors conclude that community organizing strategies at the workplace may be appropriate to reach minority rural residents. "A low-intensity community organizing approach with minimal intervention resources can reach employees in such work sites and produce small behavioral and attitudinal changes."

Health Promotion Programs by "Predominantly Rural" North Carolina Hospitals[5]

Christine Dorresteyn-Stevens researched health promotion programs offered by North Carolina hospitals with fewer than 100 beds serving predominantly rural populations. Twenty-nine of the 45 such hospitals in the State responded to her survey.

Ninety-three percent of the responding hospitals offered at least one health promotion program. At least half of the responding rural hospitals offered information on first aid and cardiopulmonary resuscitation, AIDS education, nutrition, prenatal education, and breast self-examination. Other common programs included smoking cessation, weight control, and stress management. Target audiences varied, with hospital employees being the most common, and nonpatients and members of the community ranking second. Program coordinators, 85 percent of whom were nurses, used a variety of financing methods and developed programs targeting hospital employees for certification, to reduce absenteeism, or to increase productivity. Regular staff conducted most program sessions. Having nurses or in-service departments coordinating most programs increased the likelihood that these programs would be for staff rather than the community. Dorresteyn-Stevens noted that hospitals could serve as links between health care, businesses, and community agencies and be key players in establishing a group to coordinate health promotion activities.

Summary

Partnerships are essential to reduce the community fire problem. One person or a single organization cannot possibly reduce a fire problem alone. Fire departments generally rely on mutual aid from other organizations for fire suppression efforts—apply the same strategy to community fire safety education programs.

References

1. "USDA Rural Development Housing Programs." Accessed online on May 5, 2005 at *http://www.rurdev.usda.gov/rhs/*

2. Fire Service Training Services, Alaska State Fire Marshal's Office. "Project Code Red." Accessed online *http://www.dps.state.ak.us/fire/asp/pcr.asp* on June 1, 2005.

3. Boddie, Stephanie C. "Fruitful Partnerships in a Rural African-American Community: Important Lessons for Faith-Based Initiatives." *Journal of Applied Behavioral Science 38*, 3 (Sept. 2002): 317-333.

4. Fries, Elizabeth A., Jennifer S. Ripley, Melissa I. Figueiredo, and Beti Thompson. "Can Community Organization Strategies be Used to Implement Smoking and Dietary Changes in a Rural Manufacturing Work Site?" *Journal of Rural Health 15*, 4, (Fall 1999): 413-420.

5. Dorresteyn-Stevens, Christine. "The Rural Hospital as a Provider of Health Promotion Programs." *Journal of Rural Health 9*, 1, (Winter 1993): 63-67.

Step 3: *Create a Strategy to Solve the Problem*

Creating a strategy is the beginning of the detailed work necessary for developing a successful fire safety education program. In your strategy, include what will be done, where and how it will be implemented, and who will conduct which parts of the program. Also include an evaluation plan to measure effectiveness. Creating a strategy requires carefully thinking out a plan of action that is developed through a group effort.

Creating a strategy involves the following activities:

1. Review information about the community fire problem.

2. Identify target groups.

3. Select or create your program.

4. Identify required resources.

5. Develop an evaluation strategy.

In Step 2, you developed a community planning team, which should be comprised of a core group with a primary stake in reducing the fire problem, or that can offer resources. The planning team can include people from a variety of organizations and backgrounds.

Small Town Strategy

After completing a community fire analysis, Small Town fire safety professionals determined that the highest priority fire problems were fires involving gas wall heaters and a lack of working smoke alarms, both in older homes occupied by older adults. Strategies to address these problems may involve education of homeowners in safer behaviors they can follow with respect to their heaters and their smoke alarms. Strategies also may involve a need for personnel, resources or special expertise to support acquisition, installation, or maintenance of new heaters or new smoke alarms.

Small Town has an active senior citizen organization and many churches with senior citizen programs and activities. In addition, Small Town's community development director works closely with senior initiatives.

Figure 5.1 shows the members of the planning team. Since she has experience running meetings, the community development director will lead the first meeting.

The planning team includes the following groups or stakeholders:

continued on next page

- *member of the group the fire problem affects—senior citizens;*
- *groups with a financial interest in the problem—insurance agents, heating contractor;*
- *individuals with influence—church ministers and the mayor;*
- *agencies that can reach the target group with important fire safety information—senior center, local newspaper, AARP, and ministers;*
- *organizations that can provide work (personnel) and possible wealth (funds, new equipment for installation) during the implementation of the program—church ministers, AARP, senior center, and the fire department; and*
- *individuals who can provide wisdom (including special technical expertise if needed) when developing the program—extension agent, heating contractor, newspaper editor, and insurance agent.*

Figure 4

Community Planning Team

- Fire Department
- Local Newspaper Editor
- Senior Center
- Extension Agent
- Local Senior Citizens
- Church Ministers
- Mayor
- Insurance Agent
- Local AARP President
- Heating Contractor

1. Review Information about the Community Fire Problem

The planning team must review the community fire profile and the problem statement, including the sources of the information, so that everyone on the team has a clear understanding of the community and the fire problem. Now develop a program goal in the form of a broad statement about the problem, and what the team wants to accomplish.

Small Town Program Goal

Problem: *A large share of residential fires and related losses in Small Town involve gas wall heaters in older homes of older adults who also often do not have working smoke alarms.*

Goal: *Decrease the number of residential fires involving wall heaters by 75 percent over 3 years and install smoke alarms in 30 homes of older residents each year for the next 3 years.*

Once you have developed a program goal, consider the circumstances of the specific fire problem you are targeting. Remember that fires don't simply happen—fires generally involve some type of human behavior, which may be a usage behavior by the occupant, a design or installation error by the professionals who originally installed the heater, or a maintenance error by either or both occupant and professional. Understanding the behaviors or actions that contribute to the fire problem is part of finding the best solution.

Consider the wall heater problem in Small Town. A usage error might be moving a bed against the heater. An installation error might be installing the heater without enough clearance between the exposed heating elements and the installed insulation in the walls. A maintenance error might arise because an occupant could not afford to have the heater serviced at the beginning of the heating season.

You may not have the experience needed to determine the specific actions that occur during a fire. A good source of common contributing behaviors is the narrative section of the incident report or the fire investigation report. Another good source of common types of fires is available in the reports from NFPA and the USFA (see Section 3).

---

2. Identify Target Groups

In Step 1, you started to identify the target groups, but now you should pinpoint exactly who the fire safety education program will target. A target group is a group you want to reach with important fire safety messages. In the Small Town scenario, for example, the senior residents with gas wall heaters in their homes make up one target group. The messages for this group concern keeping combustibles away from the wall heater and the importance of working smoke alarms. Another target group in the Small Town scenario may be home health care workers who frequently visit senior citizens in their homes. You may give them information about spotting gas wall heaters in need of service or that pose a fire risk. Carefully identifying who the fire problem affects most often and where is critical.

Remember to appreciate that the people from whom you are asking advice will offer valuable information about what may or may not work. Listen to them! Failure to consider suggestions from the community may result in creating a fire safety education program that the target groups reject.

Understand the needs and characteristics of the target group by speaking with them to gain insight into how to design and conduct your fire safety education efforts. Consider home health care workers, for example. By visiting with them you would find out how often they visit senior citizens in their homes, what they do while in the home, their ability to inspect the wall heaters, and their willingness to share important fire safety information with the seniors. In turn, they may have some advice for you.

If you can't to visit with the target group directly, consider visiting with others who interact with the target group. Again, consider the Small Town example. The local fire chief or fire safety educator may not be able to visit with many of the senior citizens. He or she could, however, visit with local ministers who interact frequently with the seniors.

When you visit with the target groups, remember that people generally do not like having things done to them and prefer things being done with them. Getting people to accept new ideas and change behaviors can be challenging. Working with the target groups from the beginning will help them accept the fire safety education program.

Successful fire safety education programs have target groups whose members:

- are aware of the fire problem;

- understand the fire problem and what contributes to it;

- believe that they, or their loved ones, are personally at risk;

- believe that the fire problem is unacceptable and serious;

- understand that a solution exists;

- believe they can reduce the fire problem;

- are involved with the planning process from the beginning; and

- have an opportunity to provide input and suggestions.

3. Select or Create your Program

Usually, selecting an educational program already developed by another group to address the same fire problem you are tackling is the easiest way to go. Developing a program from scratch requires a lot of time, money, and expertise—three resources you may not have. Fortunately, numerous educational programs are available to address just about every fire problem.

Successful fire safety education programs share some essential components, including:

- A presentation for delivery to the target group. The presentation may involve only a speaker providing basic information, or it may include digital slides, videos, and so on to enhance the presentation.

- Printed handout materials that support the presentation. If the handout materials, or support materials, can be used in lieu of a presentation—even better. The information in the handouts must be attractive, easy to read and understand, and must be consistent with the information in the presentation.

- Information for the media. The program should include news releases that you can give to the media. The news releases are prepared stories that provide the target group with information on the fire problem and the actions people should take. Some educational programs also provide prerecorded public service announcements.

- An evaluation tool. The evaluation tool may be as simple as a survey that people who attend a presentation can complete. For programs in schools, the evaluation tool should measure the amount of learning that occurs.

Research any existing fire safety education programs. First, contact your State Fire Marshal, who may recommend a specific, proven program. The Internet is another excellent tool. Type "fire safety education" into any search engine and add specific phrases to search those results. Consider the Small Town scenario, for example. After searching "fire safety education," adding "home heating" will reveal different resources.

Local and State agencies or organizations, such as those listed here, may have access to fire safety education programs.

- **Local County Extension Agent.** Many extension agencies have fire safety education programs dealing with rural fire problems, including some oriented to young adults involved in 4-H, and to farmers and ranchers.

- **Local and State Health Department.** In recent years health departments have begun tackling fire problems, especially in rural areas without full-time fire safety educators. CDC programs also are available through local health departments.

- **Insurance companies.** Several of the large regional and national insurance companies have fire safety education programs. Some target young children, while others target home owners.

- **State Fire Marshal's Office.** Your State Fire Marshal should offer programs and other resources, such as funding and support materials.

- **State Forestry Department.** State Forestry Departments may have education programs to deal with wildland fire problems. Also, some State Forestry Departments have rural fire coordinators who can provide technical assistance.

National agencies and organizations offer numerous programs. While you must buy some of the programs, others are available at no cost. Most of these fire safety education programs are well designed, and have been evaluated for effectiveness.

- USFA. The USFA has many programs available at no cost, as well as a variety of support materials, such as brochures, videos, and reports. Order these materials from the USFA Web site, *www.usfa.dhs.gov*

- NFPA. NFPA is another excellent source for fire safety education programs. In fact, NFPA produces two of the most widely used fire safety education programs— Learn Not to Burn® and Risk Watch®. NFPA also offers extensive support materials, such as videos, brochures, news releases, and fact sheets. Review these materials in the education section of the NFPA Web site, *www.nfpa.org*

- CDC. The CDC now provides fire safety education materials. In most States, the CDC provides smoke alarms through State health departments, as well. Get more information at *www.cdc.gov/health/fire.htm*

- Home Safety Council (HSC). HSC is a nonprofit organization that provides information and materials on home safety issues. HSC also sponsors an expert network of successful fire safety educators. Find this information at *www.hsc.org*

- National Volunteer Fire Council (NVFC). The NVFC promotes the mission of the volunteer fire service in the United States and recognizes fire safety educators annually. Get more information on NVFC programs and services at *www.nvfc.org*

- Safe Kids USA. Formerly known as National Safe Kids, Safe Kids USA is dedicated to eliminating childhood injuries, including those caused by fire. Safe Kids USA works through State and local coalitions to help fire safety education programs. Learn more at *www.usa.safekids.org*

Many other organizations and agencies offer fire safety education programs and materials, most of which are high-quality and well developed. The planning team should ask several questions to determine whether to use a fire safety education program and the supporting materials in your community:

- How well does the program address your fire problem specifically? If your fire problem involves children playing with matches, for example, but the program only briefly mentions child fireplay, the program is not suitable.

- How appropriate is the fire safety education program for your community? Does the program fit your community's unique characteristics? Was it, for example, designed for a metropolitan area, while you represent a rural area?

- How appropriate is the fire safety education program for your target groups? Consider whether the program uses terminology that your target group will understand, and if the program is appropriate for their physical abilities.

- Is the fire safety education program feasible, considering your available resources? If the program costs several hundred dollars, but you have no funding, it might not be feasible. Also, consider the number of people needed to implement it.

- Is the organization that developed the fire safety education program actively involved in fire safety? The best programs typically come from organizations whose primary mission is to prevent fires.

- Does the organization readily offer support? Some organizations provide assistance with the implementation and use of the program, which can be very important when you have minimal experience with fire safety education programs.

Once your planning team has reviewed the potential fire safety education programs, it must make a decision. If the team's first choice is a program it must purchase, try raising the money before selecting an alternative program. Don't give up because you don't have the money readily available. A common trap when selecting fire safety education programs is simply choosing the free program. Free does not equal appropriate. Many free programs are of poor quality.

If, after reviewing the available programs, your planning team is unable to find something suitable, it may need to consider developing a program. Because of the time and cost involved, however, this should be the last-resort option. If you do choose to develop a program, seek assistance from a university or organization with experience developing community-based safety programs.

4. Identify Required Resources

Through an organized process, the planning team has selected a fire safety education program. Now determine the resources required to implement the program. Some programs identify the needed resources, but usually the planning team must do so.

Some common needed resources are

- money to purchase the program, support materials, and other equipment, such as smoke alarms;

- volunteers to deliver the presentation, install smoke alarms, and so on;

- a place to hold the presentations; and

- printing of materials and notices about presentations.

5. Develop an Evaluation Strategy

The planning team's work is not quite complete, as it must develop an evaluation strategy based on the goal of the program. Use the evaluation to determine if you've achieved your goal. The evaluation plan includes the problem statement, the goal, and the actual method of evaluation. To conduct an evaluation, you must know the extent of the fire problem before the program, which establishes a baseline. You also must have an accurate record keeping system that will track the incidents of fire, the presentations delivered, smoke alarms installed, and so on.

Evaluating your program is important in many ways. Most grants available for fire safety education initiatives require some type of evaluation. Also, being able to prove past successes helps to obtain resources in the future. Finally, the community deserves to know the status of its fire safety education program.

In the Small Town scenario, for example, the baseline for fires involving wall heaters is 11 fires over 3 years. Another baseline is the number of homes—25 percent—with working smoke alarms. Next, look at the Small Town planning team's goal statement, or what it wants to achieve. The program's goal is to reduce the number of wall heater fires by 75 percent over the next 3 years, which will translate to only three such fires. Another goal is to install smoke alarms in 90 homes during that same time period.

Now, decide how the planning team will measure progress towards the goal. In Small Town, program administrators must do two things: First, they need a record of the number of fires and their causes, which the fire department already is tracking. Second, they need more detailed records on the number of smoke alarms installed in the homes of senior citizens. The more organizations involved with installing the smoke alarms the more challenging this task will be. Finally, the Small Town program administrators must monitor the records during the 3 years to track the program's progress. Monitoring the progress is essential to help the planning team determine if it needs to make changes. Total the records at the end of the 3 years to determine the program's success—or lack of success.

Turning a Tragedy into an Opportunity

In 2002, six children died in a manufactured home fire in Tchula, Mississippi. Because the home's electricity did not work, the children slept in the living room and used a candle for light. That candle started the fire. Tchula, with a population of only 2,332 people, is in Holmes County, one of the most impoverished counties in the Nation.

As a result, local leaders formed the Mississippi High-Risk Fire Safety Task Force, made up of local partner groups, such as the Community Cultural Center, the local health center, and the schools. The task force worked with local organizations, the schools, and volunteer fire departments from all the towns in Holmes County. It also partnered with two national organizations—the NFPA and the USFA. The group's goal was to install 10-year lithium battery smoke alarms in every home in Holmes County. Firefighters and volunteers met their goal, installing 8,700 smoke alarms in just over 2 years. They also taught fire safety education in the schools, and the

continued on next page

community got involved with Fire Prevention Week activities. The local leader and project coordinator, Margaret Wilson, trained the volunteers to install the alarms correctly and to provide fire prevention and escape messages to all the residents.

The effort took a great deal of organization, including phone calls, home visits, community meetings, publicity in the local media, and fliers posted in places of worship. The task force thanked volunteers with certificates, Wal-Mart gift certificates, and a recognition dinner. Wilson says that having reliable volunteers was the key to the program's success.

Your evaluation strategy does not need to be complicated. In fact, simpler usually is better. Detailed evaluation may sound easy during a planning meeting but making it happen is another thing. Detailed evaluations require significant time and record-keeping. If your planning team does not have any experience evaluating a community-based program, seek help. Usually the staff of the State Extension Service or the State Health Department includes professionals specifically trained in evaluation. When you ask for help, emphasize that you want an evaluation process that is reasonable and simple to implement.

Summary

Creating a strategy involves detailed work and a team approach to be successful. The strategy must include what will be done, where it will be done, how it will be done, and who will do it. An evaluation component measures the effectiveness of the program.

Step 4: *Implement the Strategy in the Community*

Implementing the strategy involves putting the solutions developed in Step 3 into action. In a sense, Step 4 is where the "rubber meets the road" for your fire safety education program. In the previous steps, you've already determined, or at least discussed, much of what is included in the implementation strategy.

The implementation plan provides the following details:

- how the program will be implemented, including who will do the work and when;

- the roles and responsibilities of each team member;

- checklists that identify the needed implementation steps; and

- any potential problems.

An action plan is a step-by-step outline of what you'll need to get the program started. Sometimes a chart (see Figure 5) can help plan the implementation and monitor progress.

Figure 5. Action Planning Chart

ACTION PLAN				
Date				
Program Goal:				
Step #	**Action to Be Taken**	**Assigned To**	**Resources Needed**	**Date Completed**

The actions in the implementation step are

1. Establish responsibilities and a timetable of activities.

2. Market the program to the community.

3. Start the program activities.

4. Monitor the program's progress.

5. Periodically report the program's progress.

The planning team and the members of the target groups develop the implementation plan. Often, the planning team is able to develop the plan without any additional assistance or membership. Resist any temptation to put the program into action without taking time to develop an implementation plan. Identifying who is responsible for each task is an important part of the plan. Those involved in implementing the program must be clear on what the team expects of them. A common problem among fire safety education programs is confusion about who is responsible for what, and when it should be done. Your implementation plan will reduce or eliminate this problem.

Remember, you'll need many different tasks done to make the fire safety education program a reality. You'll need some people, for example, to deliver the actual fire safety education presentations to the community and others to track the activity, do the reports, and so on. One important role of the planning team is to match peoples' skills with the needed tasks.

1. Establish Responsibilities and a Timetable of Activities

You must coordinate the implementation of the program with team members completing the tasks in the proper order. To do this, the planning team should develop a predelivery checklist of everything that must be done before fully delivering the fire safety education program. Think of the predelivery checklist as your checklist for success.

Figure 6. Sample Checklist

> **Small Town Fire Safety Education Program Predelivery Checklist**
>
> ☐ *Recruit Small Town volunteer firefighters to deliver fire safety education presentations.*
>
> ☐ *Schedule fire safety education presentations at the senior center.*
>
> ☐ *Schedule presentations at AARP and Lions Club meetings.*
>
> ☐ *Make copies of the handout materials.*
>
> ☐ *Send a notice regarding the presentations to the local churches and to the local newspaper.*

In order of completion, the checklist identifies

- the people you should notify about the implementation of the fire safety education program, including team members, target groups, and any other interested parties;

- the equipment and materials needed, which you identified in step 3; and

- appointments and meetings you need to schedule, including fire safety education presentations that are part of the program.

The checklist is a road map for the tasks that you must complete. It is not, however, foolproof. Expect additional needs to come up as the program gets underway.

Rarely does anyone implement a community fire safety education program without any problems. Some problems, however, are pretty common and can be expected. Have a backup plan for these anticipated problems.

To identify potential problems, bring the program team together with representatives from the target groups. Once you've assembled the group, have an open discussion about the potential problems. Brainstorm problems and feasible solutions. Involving team members who will be directly involved in the implementation of the fire safety education program is especially important, as their experience will be invaluable.

2. Market the Program to the Community

The program can work only if the community, particularly the target group, is aware of it and its purpose. Make sure you keep the community informed of the program and its progress. Many people in the community will be interested in the program, including members of the planning team, elected officials, sponsors, fellow firefighters, members of the target groups, the general public, and members of the program team involved in the implementation.

The program's marketing information should inform people about the effectiveness of the program in achieving its goal of reducing fire in the community. In other words, you want to let people know the program works! Any future fire safety education program may depend on how well you market the results of this program. When providing information to the public, always use familiar, easy-to-understand terms. Have someone who was not involved in developing the program review the information for clarity. Make any needed changes before releasing the information.

Communicate the following information to the community:

- the details about the program's activities, including the number of presentations, the time and location of the presentations, and so on;

- the goal of the program;

- the organizations and individuals who helped develop and implement the program;

- the reasons for the program;

- any anecdotal information about those involved in the program, including members of the target groups; and

- any evidence that the program is working to reduce the number of fires.

You can use several methods to get information to the public. Most likely, no single method will be enough; use at least two simultaneously to ensure everyone gets the information. Some methods are more common than others.

Local media. Generally, your local media is the best way to reach the general public, whether through news stories, ads, or letters to the editors. Meet with a local reporter to explain the program and why you need their help. Be specific about what information you want to get out and who you want the information to reach.

If you've chosen a prepared fire safety education program with prepared news releases, use them. Providing local media with a prewritten story greatly increases your chances of getting the story printed. If necessary, make some changes to the story to make it fit your community. Also, always provide pictures when appropriate and available.

Direct mail. If the fire safety education program involves a target group in a specific part of town, a direct mail piece sent to every resident in that part of town may be effective. The local Chamber of Commerce, the city or county clerk, or a local utility may provide address lists. Keep your mailing to two pages or less, and make sure the reader can read it and understand it within a few minutes. As a rule, people will ignore or throw away anything that takes longer than 2 or 3 minutes to read.

Meetings. One of the most effective methods for communicating your information is through meetings with citizens, which may include church meetings, community groups, and so on. A meeting provides you the opportunity to answer questions and emphasize the importance of the fire safety education program. Consider recruiting someone with experience in media communications to help get information to the community. Consider someone at your local newspaper, college, health department, or a local business with this kind of experience and training.

Some tasks focus on the team rather than the public. One such task is the recognition of team members. Successful fire safety education programs are the result of hard work by people committed to improving the community. These people are willing to sacrifice their time and expertise for the program's goal.

Recognize anyone involved in developing and implementing the program, giving the highest recognition to those who contribute the most. Ways to recognize the efforts of the team include certificates of appreciation, gift certificates to local businesses, items that can be used at home or work, letters of appreciation by local elected officials, and appreciation dinners or other community events. If possible, plan your recognition of team members at a public or community event where a large segment of the community gathers, such as an annual dinner, a parade, or a sporting event. The large community presence gives extra meaning to the team members and shows the community the importance of the program.

Regardless of the recognition method, make your recognition sincere and meaningful and send the message that you and the community valued the sacrifice and hard work. Meaningful recognition emphasizes the importance of the contribution and the program's mission. If you do not recognize team members, they may feel you've taken advantage of them, and that the community as a whole will minimize their efforts.

3. Start the Program's Activities

When you and the team have developed the final program, put everything the team has worked on to use in the implementation. Follow the program implementation plan to avoid any confusion or problems. Use the action plan to guide the implementation. Also, follow through with responsibilities and schedules.

Communication among the team members is critical to the implementation. Keep team members informed on the progress of the implementation, as well as any issues that arise. Personal face-to-face communication is the best, but e-mail will work. Also, consider periodic meetings to discuss the program to provide the team an opportunity to address any problems and come up with appropriate solutions.

4. Monitor the Program's Progress

Remember to monitor the progress of your fire safety education program by following the action plan. Everything needed to monitor the program should be in place *before* you implement the fire safety education program. Following through with commitments and responsibilities is sometimes all you'll need.

To monitor progress, visit with team members actually doing the program, who can give you timely, direct feedback, as well as anecdotes helpful for the evaluation and for marketing the program to the community.

5. Periodically Report the Program's Progress

Throughout the implementation of your program, periodically communicate progress to the program team and the community. The program team includes not only those working directly on the program, but also any elected officials, appointed officials, and community leaders who supported developing the program. The community includes the target groups, as well as the community at large.

Direct communication, including verbal updates at meetings, is the most effective way to report progress to the program team. When verbal updates are not possible or timely, use written or e-mail communication, keeping it concise, succinct, and no more than one page.

The local media is a great outlet for communicating with the community and target groups. Again, keep your information concise and succinct and provide an update only when you have something worthy to report. Too-frequent reports lose their impact. In a sense, your progress report should be worthy of a headline. Report the information of interest to the program team and to the community, including:

- human interest stories of how the program prevented a fire or injury to a local family;

- completion of major benchmarks in the program, such as mounting 500 smoke alarms in the first month;

- the completion of a program goal or objective;

- the accomplishments of the program at the end of a chosen period, such as the 2-year anniversary;

- award of a major grant or other resource; and

- recognition of the program by a government agency, such as the State Fire Marshal.

Remember, the information must be timely—old news will have no importance.

Finally, monitoring the progress of your program may mean revising the program. If you are not making progress towards your goals, you and the team may have to revise the original plan.

For more information and tips on marketing your fire safety education program and your fire department to the community, order a copy of the USFA manual Strategies for Marketing Your Fire Department Today and Beyond *at no cost through the USFA's Web site* www.usfa.dhs.gov

---

Summary

Implementing the strategy involves putting the plan into action in the community. Sometimes, you'll need to revise the program for continued progress. Marketing the program to the community is an important part of implementation that you should not overlook. Monitoring the progress of the program and preparing regular progress reports are important, as well.

Step 5: *Evaluate the Results*

The primary goal of the evaluation process is to demonstrate that your fire safety education program is reaching the target groups in the community and is achieving the desired results. In other words, evaluation is all about determining whether you achieved the program's goal. There is much at stake in the evaluation process, and because of that it is very important and must be given appropriate effort.

Often, people overlook, or intentionally skip, the evaluation step. Fire departments generally give three reasons for not evaluating fire safety education programs—fear of working with statistics, fear that a good evaluation may identify problems in the program, and lack of evaluation experience.

Many people believe that evaluation involves difficult, complex math formulas and calculations, which is simply not true. Evaluation involves identifying the current problem, determining the future fire loss you are working towards, making some comparisons, and reaching some conclusions about the results of the program. Simple math is more than adequate.

Many fire departments and planning teams don't know how to evaluate a fire safety education program. Fortunately, help is readily available from numerous sources, including the USFA, the NFPA, local teachers and college instructors, public health officials;, extension agents, mental health officials, and the State Fire Marshal's Office. Available resources include evaluation manuals and handbooks, reports of successful programs, and staff members who can help design an evaluation strategy.

A good evaluation process will prove that your program achieved its goal. Failure to evaluate may result in a less-than-successful program. Seeking help from others who regularly conduct evaluations will be a wise investment.

The actions in the evaluation step are:

1. Collect program data.

2. Compare data.

3. Modify the program when needed.

4. Report program results.

Evaluation involves comparing the community's fire problem after the program with the fire problem before the program.

> *For evaluation to be effective you must have accurate information on the local fire problem. You should have addressed this issue in Step 1. The old saying, "garbage in, garbage out" certainly applies to the quality of your evaluation process. Before proceeding it may be helpful to review the information in Step 1: Conduct a Community Analysis.*

Evaluate two areas, the first of which is the program's activity, which includes the amount of work done by the team members. In the Small Town scenario, for example, the team had a goal to install smoke alarms in 30 homes each year, which is a program activity. Other activities include the number of fire safety education presentations delivered, community events attended, and news articles published.

Keep in mind that activities don't always reduce the fire problem. Generally, well-designed activities will reduce the fire problem. Sometimes, though, the activities do not reduce the fire problem. In short, don't assume that doing lots of things always will reduce the fire problem.

Also evaluate the impact of the program. The impact is the effect the program has on the actual fire problem. In Small Town, for example, the team had a goal to reduce the number of residential fires involving wall heaters by 75 percent over 3 years. The impact of the program will be the actual reduction of those fires after 3 years of fire safety education activities. If, during the 3 years of the program, the town experienced only two wall heater-related fires, the impact of the program was a 75 percent reduction in the number of fires. Always remember that impact measures the actual change in the fire problem the planning team identified in the community profile. You cannot determine impact without accurate data on the fire problem.

1. Collect Program Data

In Step 1, your planning team developed a community profile that described the fire problem in your community. In that process you identified a specific fire loss, which may have included the number of fires, the types of fire, the people affected by the fires, and so on. This starting-point profile information is known as baseline data, and you will use it in the evaluation process.

The profile information on your community's fire problem may include the frequency of the fires; location of the fires; the time, day, and month or the incidents; the cost in terms of economic loss, loss of life, and injuries; and the cause of the fires.

2. Compare Data

Doing your evaluation well and early can provide the framework to set up a successful fire safety education program. Establish a starting point—or baseline—and identify a destination or goal—or benchmark. You identified the baseline data in the community profile and you developed the benchmarks, or program goals, in Step 3. The benchmarks are the desired change as a result of the fire safety education program.

A review of the Small Town scenario helps illustrate this concept.

3. Modify the Program When Needed

Monitoring the progress of your program is important and is why the evaluation strategy measures both activity and impact. Following through with each can provide tangible evidence that the fire safety education program is moving towards its goal. Evaluation must be valid and objective to measure exactly what you want to measure without bias or prejudice.

Usually, impact evaluation is a long-term process and, as such, is difficult to complete before the program ends. Measuring the impact of the program on the fire problem may take three to five years. As the program progress you can, however, measure the program activity by comparing the actual program activity against the goals of the program. This comparison can provide you and your team with valuable information about the progress of the program.

If one of the program goals was to deliver 15 fire safety education presentations a year to elementary school children, for example, and you only delivered 5 in the first year, you know you need to modify the program. Maybe you need to change your presentations to the schools or maybe you need to recruit more educators. Regardless, you know that you have to change something to get the program back on track.

Small Town Scenario

Community Profile (Problem Statement): Wall heaters cause 75 percent of the fires in our community—eight fires in the past 3 years. More than half of the residential fires involved retired citizens living in older homes. Most of the unattended fires resulted from a malfunction of the heater, or combustible materials too close to the heater. Also, only one in four homes in Small Town have working smoke alarms.

The Small Town Volunteer Fire Department proposes developing a fire safety education program to address the wall heater fire problem among retired citizens. The reasons for this fire safety education program include

- *A majority of the residential fires in Small Town involve gas wall heaters.*
- *These fires affect our retired senior citizens, a group that is at highest risk from fire.*
- *Seventy-five percent of our community's homes do not have working smoking alarms.*
- *A fire safety education program has a strong potential for reducing the fire problem involving senior citizens in Small Town.*

continued on next page

The eight fires involving wall heaters in the past 3 years, and the mere 25 percent of the homes in Small Town having smoke alarms constitute the baseline. The program's goal is to decrease the number of residential fires involving wall heaters by 75 percent over 3 years and install smoke alarms in 30 homes of older residents each year for the next 3 years.

Reducing residential fires involving wall heaters by 75 percent over 3 years and installing smoke alarms in at least 30 homes of older residents each year for the next 3 years are the benchmarks.

Small Town can use its fire reports to measure the impact of the fire safety education program. The fire reports will identify the cause of any residential fire.

Small Town can use program reports to measure program activity by examining the number of smoke alarms installed each year. Team members installing the alarms must, however, keep accurate reports, including identifying that the home was that of an older resident.

Of course, if the end-of-program evaluation results show that you did not achieve your goals, the planning team will have to figure out why. The team should be able to identify what to change and incorporate those changes into the next program or in the modification of the current program.

4. Report Program Results

Share results of the evaluation process with the planning team, your fire department, the target groups, political leaders and decisionmakers, and the community in general. Once your team has developed some conclusions from the evaluation, provide that information to the planning team. In fact, it is best to hold a team meeting to review and discuss the results.

At this point you and your team should have decided whether you need to modify the fire safety education program. If the team decides it must modify the program, inform everyone involved of the decision and the reasons for the modification.

Once you've completed the program, report the results of the program to the target audience, the organizations involved, and the community. Use the conclusions from the evaluation process as the basis for the report, and let them determine the report's length.

Generally, the program report should include

- the problem statement;
- the community profile;
- the program's goals;
- an overview of the program, including partners and supporting organizations;
- a summary of the activities undertaken during the program;

- highlights of the program, including anecdotal stories;

- the impact of the program in terms of the fire problem;

- a summary of the conclusions; and

- any recommendations.

Summary

Some steps are essential to perform a successful evaluation. First, you and your team must commit to the evaluation knowing that the process will take time and effort. Then, if you need help, ask someone with experience in performing evaluations. You must complete the evaluation as you designed it. Keep an open mind and be prepared to make changes based on the findings of your evaluation.

Summary

Many fire departments in small communities shy away from starting fire safety education programs due to a lack of resources and expertise in fire prevention. Every small community has the ability to develop and implement a successful fire safety education program to reduce its fire problem. This initiative requires a commitment by department and community leaders, as well as the community working together to solve the fire problem.

Successful programs usually follow a five-step process:.

Step 1: Conduct a Community Analysis.

Step 2: Develop Community Partnerships.

Step 3: Create a Strategy to Solve the Problem.

Step 4: Implement the Strategy in the Community.

Step 5: Evaluate the Results.

Working as a community to follow the five-step process makes identifying and reducing the local fire problem possible. This process will result in a safer community and a stronger relationship between the fire department and the citizens of the community.

www.ingramcontent.com/pod-product-compliance
Lightning Source LLC
Chambersburg PA
CBHW081358170526
45166CB00010B/3132